Maria Aparecida da Silva Alves
Fernando Salomão
Ibraim Fantim

Environmental characterisation of the springs

Maria Aparecida da Silva Alves
Fernando Salomão
Ibraim Fantim

Environmental characterisation of the springs

In APP areas

ScienciaScripts

SUMMARY

CHAPTER 1

INTRODUCTION

This book was motivated by the large increase in population, linked to the accelerated and unplanned process of urbanisation, which has put great pressure on natural resources in Chapada dos Guimarães. Its object area - *the* Quineira stream basin. Based on the hypothesis that by identifying its springs, characterising them and diagnosing the vegetation cover, erosion processes, housing forms and occurrences, and the quality of surface water resources, it will be possible to propose actions that enable environmental preservation, minimising the effects of degradation and even enabling the recovery of degraded areas. Water is the source of life, all living beings depend on it to live. However, as important as it is, water bodies and springs are being destroyed.

Water is probably the only natural resource that has to do with every aspect of human civilisation, from agricultural and industrial development to the cultural and religious values ingrained in society. It is an essential natural resource, whether as a biochemical component of living beings, as a means of life for various plant and animal species, as a representative element of social and cultural values and even as a factor in the production of various final and intermediate consumer goods (BRUCE, 1993).

According to Barros et. al (2007), the Earth is largely made up of water, 70% of its surface is covered by this essential liquid for life, making it one of the most abundant resources on the planet. However, of all the water that exists, only a small portion, fresh water, can be used for human consumption, after adjusting its physical, chemical and biological characteristics, making it drinkable. Tundisi (2004) states that the supply of good quality fresh water is essential for economic development, for the quality of life of human populations and for the sustainability of the planet's cycles.

According to the Brazilian Forest Code (BRASIL, 1965), Permanent Preservation Areas (APP) have the environmental function of preserving water resources, the landscape, geological stability, biodiversity, the gene flow of fauna and flora, protecting the soil and ensuring the well-being of human populations. However, the occupation of these areas by urban populations is becoming increasingly common, as they settle without any planning and cause the degradation of the APP, and of the

springs. According to Calheiros et al. (2009), springs provide good quality, abundant and continuous water, located close to the place of use and at a high topographic level, making it possible to distribute it by gravity, without wasting energy.

Springs, watercourses and reservoirs, although different from each other in terms of their preservation strategies, have the following basic points in common: control of soil erosion by means of physical structures and plant barriers, minimisation of chemical and biological contamination and actions to mitigate water loss through evaporation and consumption by plants (Valente 2005).

Regardless of its origin or denomination, the vegetation that borders springs and watercourses is fundamental for environmental preservation, especially for maintaining water sources and biodiversity (STRECK, 2007). Disordered urbanisation in river basins generates various environmental imbalances and, consequently, damage to humans. One of the consequences of unplanned land use and occupation is the alteration of certain processes inherent to the hydrological cycle, as well as the reduction of vegetation cover and soil sealing (TUCCI, 2003).

TUNDISI (1999) states that changes in the quantity, distribution and quality of water resources threaten human survival and that of other species on the planet. In the urban area of Chapada dos Guimarães MT, this reality is also frequently observed. Some homes are located in permanent preservation areas associated with three watercourses such as the Quineira, Monjolo and Samambaia streams, the latter being the main source of water for the population of the city of Chapada.

In the Quineira stream micro-basin there is great concern about the preservation of the springs, as there are many buildings and homes around them, and more and more of these homes are taking over the stream, which can lead to environmental imbalance. The large production of domestic and construction waste can cause these springs to disappear.

The diagnosis of the Quineira stream micro-basin was carried out using the approach known as VERAH (Vegetation, Erosion, Waste, Water and Housing), a methodology developed by OLIVEIRA (2008). This approach provides insight into the environmental problems of urban watersheds, as well as an environmental diagnosis of the study area.

CHAPTER 2

LITERATURE REVIEW

The town of Chapada dos Guimarães is located 65 kilometres from Cuiabá, the capital of the state of Mato Grosso. It has a population of 17,377 inhabitants, 9,877 of whom live in the urban area, although the urban population increases considerably at weekends due to tourism (SCHREINER, 2009). The Córrego Quineira watershed is located in the town of Chapada dos Guimarães (MT), part of the Cuiabá river basin and the great basin of the Paraguay river, with an average altitude of 800 metres, situated on the plateau of the Guimarães Plateau. Within this watershed is the Quineira Municipal Park, established by Law 8.615 of 26 December 2006. Figure 01 shows the delimitation of the Quineira stream, its watercourse, its vegetation is denser in a small part of the stream, and around it is clearly visible the large amount of housing, most of which is irregular, which compromises the preservation of the micro-basin as a whole, especially the functioning of the springs and its watercourse.The main aim of creating this park was to protect the springs in the area, since the Quineira stream is used to collect water for public supply, according to Law No. 8,615 of 26 December 2006. According to the SAA (Chapada dos Guimarães Water Supply Service), the Córrego Quineira micro-basin has an area of 3.43 km^2 , its main river, the Cachoeirinha River, is 2,688 metres long and the compactness coefficient calculated for the area was 1.25, indicating that the shape of the basin is elongated and therefore not very prone to flooding. The Quineira stream is a tributary of the Monjolo stream, which in turn is a tributary of the Cachoeirinha river. This river joins the Lagoinha River, giving rise to the Quilombo River, which flows into the Casca River, a tributary of the Manso River. The Manso River is a tributary of the Cuiabá River, which contributes to the Paraguay River, the main source of the Paraguay basin, which is largely responsible for maintaining the Pantanal. The vegetation of the Quineira watershed is of the Savannah type, represented predominantly by cerrado, with the presence of riparian and gallery forest.

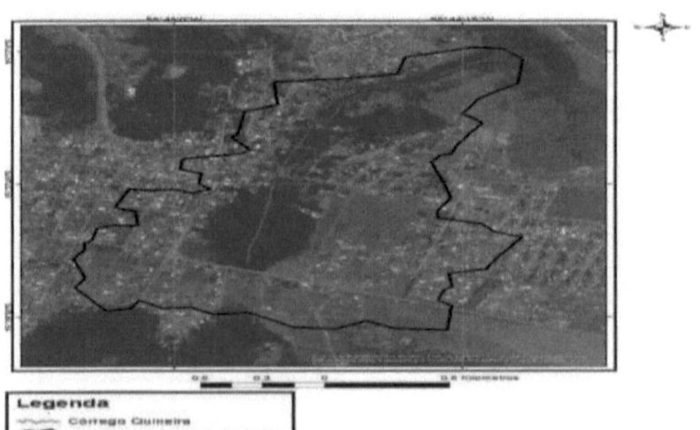

Figure 01: Delimitation map of the municipality of Chapada dos Guimarães - MT.
Source: The author (2013).

Its geological formation is characterised by sedimentary rocks from the Fumas and Ponta Grossa formations, partially covered by detrital-lateritic Tertiary/Quatemary sediments. Shallow soils predominate, with the occurrence of sub-flourishing rock and ferruginous armour. In some places, there are deep soils (Latosols and shallow Quartzarenic Neosols) and plinth soils. The slopes of the Córrego Quineira are dominated by Plintossolo, which is a poorly permeable soil. The rain that falls on this soil partially infiltrates and is barred by the armour, flowing away in an instant and forming the springs (SALOMÃO 2007).According to Maitelli (2005), according to Strahler's classification, the south-central portion of the state of Mato Grosso has a tropical climate, with a high concentration of rainfall during one half of the year (October to March), and a decrease in rainfall in the other half, from April to September.Due to the distinct topographical and geomorphological characteristics, it is possible to recognise three relief compartments in the Chapada dos Guimarães region: the Chapada dos Guimarães geomorphological sub-unit, which develops predominantly on the rocks of the Fumas, Botucatu and Ponta Grossa Formations and has elevations ranging from 600 to 800m; and the Plateau (Radambrasil, 1982). The Quineira Municipal Park is part of an urban conservation unit created to protect the springs of the Prainha stream, the object area of which is within the Chapada dos Guimarães Municipal Park. It has 26 hectares of gallery forest, with a maximum canopy height of 20 metres. This conservation unit is surrounded by pastures, houses and a small area of 2 ha of Cerrado. The urbanised area of Chapada dos Guimarães was the subject of a study aimed at drawing up a geotechnical map, summarising the elements that make up the physical environment (geological substrate, landforms and features and soil cover). They are characterised and analysed in an integrated way in order to interpret the water functioning of the slopes and determine the susceptibility of surface dynamic processes, especially erosion, mass movements and flooding (Salomão et al. 2012). Figure 02 reproduces the geotechnical map resulting from these studies, highlighting the existence of six geotechnical units in the region.

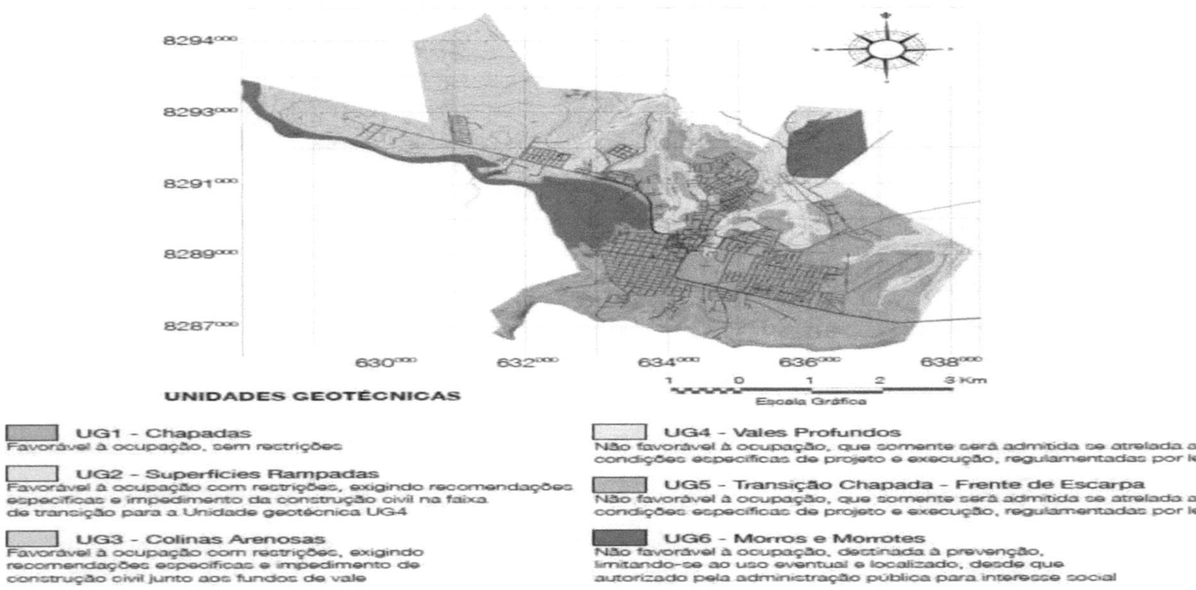

Figure 02: Geotechnical map of Chapada dos Guimarães - MT.
Source: Salomão, Madruga & Migliorini (2012).

The investigations of the soil cover and bedrock were carried out more rigorously on representative slopes, identifying and characterising the soil profiles and their respective soil horizons in relation to morphological characteristics, so as to allow interpretation of the hydrological functioning and deduction of geotechnical behaviour. According to Salomão et al. (2012), six geotechnical units stand out in the region studied: UG1 - plateaus; UG2 - sloping surfaces; UG3 - sandy hills; UG4 - deep valleys; UG5 - plateau - escarpment transition and UG6 - hills and slopes.

In the Córrego Quineira watershed, the geotechnical units UG1 - plateaus and UG4 - deep valleys were detected. Both will be described, since their characteristics have a decisive influence on understanding erosion processes as a function of land use and occupation.

Geotechnical unit UG1 - Chapada is located at the top of the Guimarães Plateau, where urban occupation is currently concentrated, with slightly flattened topography, gentle ramps and a maximum slope of 6 per cent, heading towards the deep, notched valley bottoms represented by UG - 4. Geotechnical Unit UG1 is made up of Red Latosol and Red-Yellow Latosol, with a clayey texture, and Petric Plinthosol, associated with rocks belonging to the Ponta Grossas Formation and the Fumas/Ponta Grossa transition, made up of sandstones and claystones. There is ferruginous armour on the rocky substrate, and in the case of the Latosols it is found at depths of more than 2 metres, while in the Plinthosols it practically outcrops on the surface of the land. These soils probably originated from the weathering process of the ferruginous armour.

Its geotechnical attributes include a pedological cover with low compressibility and good load-bearing capacity, as well as non-expansive materials that are easy to excavate, with artificial slopes that are resistant to landslides and of good quality as compacted embankments (Salomão, 2007).

According to Salomão et al. (2012), most of the watercourses observed in the area, including the Quineira stream, have very steep valley slopes and are carved into rocks of the Fumas and Ponta Grossa formations, which have been characterised as Geotechnical Unit UG4.

The slopes and much of the valley floor have shallow to deep soils, known as Litholic Neosols and Peritic Plinthosols, as well as rocky outcrops represented by sandstones and claystones from the Fumas and Ponta Grossa formations (Castro et al, 2000). If the soil is devoid of vegetation cover, there may be gullies that destabilise the valley slopes, favouring rockfalls and slips of the surface layer of the soil cover, silting up the valley bottoms. In this way, watercourses, which are generally intermittent, can be destroyed and the contamination of water by pollutants from urban occupation can be intensified. Roads, or even tracks and/or paths that run downhill towards the valley bottoms, intensify erosion, even if the slopes are covered in vegetation.

The fragility of this Geotechnical Unit to erosion is evidenced by the occurrence of deep gullies along the valley bottoms. The main cause of these erosive events is the inaccuracy of drainage works in the urbanised area that borders the valleys, allowing rainwater with high runoff energy to overflow from the streets.

The slopes of the valleys should not be cleared as they are important means of protection against erosive events and maintaining stability against mass movements. Any road works to cross the valleys within this geotechnical unit must be carried out with the required technical rigour, giving priority to rainwater drainage works in order to discipline the runoff with controlled energy, preventing erosion on the slopes and valley bottoms.

CHAPTER 3

Bibliographical review

3.1 River Basins and Springs

According to the Geological Glossary 1 (1999), a hydrographic basin is the area of a surface water drainage system, originating from springs and/or rainfall, occupied by a river and its tributaries and limited by the ridge (interfluve) that topographically divides this area from another neighbouring drainage basin. The hydrographic basin is defined as a natural catchment area for precipitation water, which causes runoff to converge towards its outlet (SILVEIRA, 2001).

The hydrological behaviour of the basin is influenced by the type of vegetation cover as well as geomorphological characteristics such as shape, relief, geology, soil, vegetation, among others. Souza et al (2002) cites that the soil, water, vegetation and fauna components are in permanent and dynamic interaction, and natural and anthropogenic interference affects ecosystems as a whole. In this way, water resources are indicators of ecosystem conditions in relation to the effects of the balance of interactions between the respective components. Cristofoletti (1980) considers that the drainage basin is made up of various interrelated flow channels, which constitute the river drainage, and the amount of water that reaches the river courses depends on the size of the area occupied by the basin, the total precipitation of the regime and evapotranspiration and infiltration, the drainage basins can be classified according to the overall runoff in 4 categories:

- Exorheic: when the water flows continuously to the sea or ocean, i.e. when the basins flow directly into the sea; Endorheic: when the drainage is internal and does not flow to the sea, flowing into lakes or dissipating in the sands of deserts, or getting lost in karst depressions; Arreic: when there is no structuring in a hydrographic basin, as in desert areas where rainfall is negligible and dune activity is intense, obscuring the drainage lines and patterns; **Cryptorheic:** when the basins are underground, as in karst areas, the underground drainage ends up emerging in springs or integrating into subaerial rivers.

The presence of confluences at right angles in the dendritic pattern is an anomaly that should generally be attributed to tectonic phenomena. This pattern is typically developed on rocks of uniform resistance or on horizontal sedimentary structures (CRISTOFOLETTI, 1980).

According to Valente (2005), "small basins positioned at the extremities of larger basins, generally in areas of greater slope, are known as headwater basins, or simply headwater basins. They are responsible for the formation of streams, or even brooks and creeks, as they are popularly called." Every river starts somewhere. That "somewhere" is its headwaters, the network of small rivers, headwaters, that cover the landscape of the entire river basin. The health of large rivers depends above all on intact micro-basins (CALUURI and BUBEL, 2006). According to the Environmental Dictionary, a source (2004) is a place where the eyes of water give rise to a river course. Olho d'água, or spring, is understood to be a place where water appears due to an upwelling of the water table (CONAMA Resolution No. 04, of 18/09/85).

The National Water Agency has published a glossary that brings the updated concept of Nascente (spring) into line with the understanding presented by the National Environment Council in Resolution 303 of 20 March 2002. This concept makes clear the condition for a given piece of land to have a spring; the presence of underground water. "Groundwater is understood to be the concept presented by the Environmental Ecological Dictionary - the supply of fresh water beneath the surface of the earth, in an aquifer or in the ground, which forms a natural reservoir for human use." The International Hydrological Glossary defines the typology of springs as:

Spring: A place where water naturally emerges from a rock or soil onto the surface of the ground or into a body of surface water.

Contact spring: one in which water flows from a permeable formation underlying a relatively impermeable formation; Depressional or gravity spring: one that emerges onto a surface due solely to the fact that this surface intercepts the level of the aquifer; Intermittent or periodic spring: a spring whose flow occurs only in certain periods and ceases in others; Diffuse spring: one that emanates from a permeable medium onto a relatively large area.

According to Valente (2011), springs are surface manifestations of groundwater, giving rise to watercourses; each watercourse has its own spring, and are therefore surface manifestations of groundwater that result in the formation of streams. Springs are natural environmental systems in which groundwater infiltrates temporarily or perennially, forming drainage channels downstream (Felipe, 2009). However, Valente and Gomes (2005, p. 147) point out that "the spring is a natural phenomenon that transcends the point where it manifests itself, being the result of a hydrological process that takes place in a contributing area called the hydrographic basin". Depression springs can manifest themselves as well-defined bubbling points, called waterholes, or as small surface seeps spread over an area that is soaked (swamp) and accumulates water in puddles until it begins to flow continuously, being known as diffuse springs (VALENTE, 2011). Springs from artesian water can be contact springs, which usually occur in mountainous regions with steep slopes between neighbouring areas, which facilitates the upwelling of the impermeable layers responsible for the upwelling of the water table (TORRES; JUNQUEIRA, 2005).

3.2 Permanent Preservation Areas (APP)

The concept of Permanent Preservation Areas (APP) in the Brazilian Forest Code (Law 4.771 of 15/09/1965) emerges from the recognition of the importance of maintaining the vegetation of certain areas - which occupy particular portions of a property. According to the Brazilian Forest Code, Permanent Preservation Areas (APP) are areas "covered or not by native vegetation, with the environmental function of preserving water resources, the landscape, geological stability, biodiversity, the gene flow of fauna and flora, protecting the soil and ensuring the well-being of human populations". Examples of APP are the marginal areas of bodies of water (rivers, streams, lakes, reservoirs) and springs; hilltop and mountain areas, areas on steep slopes, sandbanks and mangroves, among others. The definitions and limits of APP are presented in detail in Resolution 303 of 20/03/2002, which sets out the parameters, definitions and limits of Permanent Preservation Areas for springs. Law no.[0] 12.651 of May 2012 provides for the protection of native vegetation.

According to information from the coordinator of the Quineira Water Supply and Sewerage System, *the* micro-basin is only used for urban supply, which supplies the town of Chapada dos Guimarães with water for approximately 13,500 people, but this number increases at weekends, as the number of tourists visiting the town is high, thus increasing water consumption. SAAE also reports that the daily demand is around 3 to 3.5 million litres/day, however, it is not always possible to meet the demand even using the three catchments: the Monjolo stream (120 m3/h), Quineira (60 m3/h), and Buracão (30 m3/h, used during the dry season), especially on festival days. The Quineira pumps work 24 hours a day during the rainy season (during the dry season it depends on whether there is water available), while the Monjolo pump works 21 hours a day during the rainy season and 24 hours a day during the dry season. During the dry season, rationing is carried out in the neighbourhoods, the city's water supply service hires water trucks to supply homes, and campaigns to raise awareness among the population about water rationing.

CHAPTER 4

MATERIAL AND METHOD

In order to clarify, in the most didactic way possible, the procedures and techniques used to meet the proposed objectives, the activities carried out will be highlighted below, organised into four work stages in a logical sequence.

First stage: Cartographic Delimitation of the Target Area and Delimitation of the Base Map.

This consisted of delimiting the Quineira stream micro-basin based on a topographical map, highlighting the watercourses and water dividers. Maps were also drawn up using the Arcgis 9.3 computer programme, which delimited the area under study. For field navigation, a handheld GPS (Garmin Map 72) was used, with the image map inserted into the device's base map (precise navigation), and the walking method was used throughout the area.

The analysis of the vegetation dynamics of deforestation and degeneration was carried out on a scale of 1:50,000, of the vegetation of the Quineira stream micro-basin, over a period of 20 years, using the Geographic Information System (GIS). The categories of information were selected: vegetation cover from 1980 and 2014, for all stages of image processing (geometric correction, segmentation, classification and cross-tabulation.) The vegetation cover information was compiled from the visual interpretation of Landsat -5 satellite images and colour composition, band 3(R), 4(G) and 5(B), at a spatial resolution of 30 m, and fieldwork at the time.

Second stage: Application of the VERAH method in the Quineira stream watershed. In this stage, the environmental diagnosis of the Quineira watershed was carried out using the VERAH method, with the participation of classmates from the Integrated Basin Studies I course of the Master's Programme in Water Resources. The exploratory visit took place throughout the Quineira watershed on 14 May 2013, observing each theme of the VERAH method, the vegetation, the presence of erosive processes and ways of controlling them, the presence of solid waste in the watershed, human occupation, water abstraction from the spring and the presence of springs.

Vegetation cover was identified in the micro-basin following the classification of Carvalho (2003), in accordance with Lima et. al (2011). Four (4) points were characterised along the watershed, with the source being called point 1 (Pl), the catchment point 2 (P2), the pool point 3 (P3), and the outflow point 4. Four 50 m transects were made, perpendicular to the watercourse of the Quineira Stream, sweeping 2 m to each side, totalling 4 m, with the tree species that were above 15 centimetres in diameter at breast height (DBH) from the source to the outlet: pruning shears to fit the required sample size, newspaper to separate each sample and help with the drying process in the oven, string to tie up the collection presses. All the samples were catalogued with their respective collection data, as shown in figure 3 below.

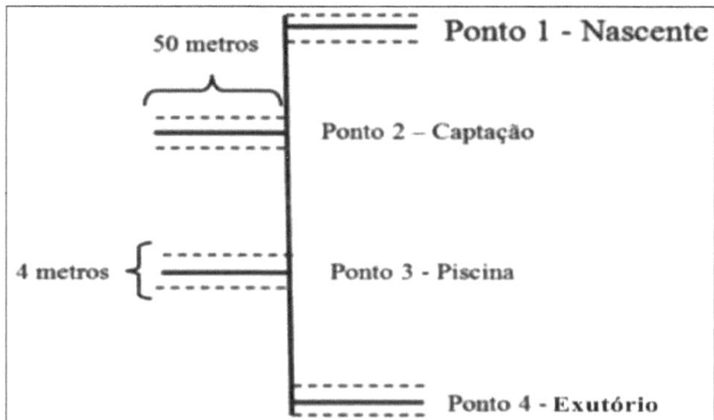

Figure 03: Layout of the transects in the Quineira stream.
Source: The author (2014).

The vegetation material collected was pressed and dried in an oven at the Biosciences Institute of the Federal University of Mato Grosso. The material was dried in a wooden oven with 60-watt bulbs, keeping the temperature between 40° and 45° C for approximately 72 hours. After this process, the botanical material was sent to the UFMT Herbarium for taxonomic identification, and bibliographies and databases were used to identify each species. The erosions present were registered on the Erosion Registration Form proposed by the Technological Research Institute of the State of São Paulo - IPT - (1990). On this form it was possible to distinguish geometric data on erosion, data on the local characteristics of the physical environment and land use, the history and dynamic evolution of the process and the control measures adopted.

Firstly, *the* erosion was identified, its location and access area. Next, *a* registration form was duly

13

filled out, in an attempt to determine the causes of the development of erosion, the susceptibility of linear erosion processes, and environmental changes, especially in relation to *the* degradation of watercourses and springs.

WASTE AND HOUSING

A socio-economic and environmental questionnaire was administered to 40 households and 23 commercial establishments. The aim was to obtain a socio-economic profile of the population and their level of awareness of the issue, as well as to gather information on the transport, frequency and final disposal of the solid waste generated. The socio-economic and environmental questionnaire was administered randomly in both the commercial and residential areas surrounding the Quineira stream micro-basin.

Surface water resources were characterised empirically, adopting procedures proposed by Oliveira et al. (2008) and Guedes (2010) in terms of the following aspects: flow, the occurrence of watercourses, colour, odour, or whether it indicates the presence of galleries and water and sewage pipes. The parameters were listed following a survey of the methods and materials used to collect and preserve samples. The bottles were separated and washed in accordance with the National Guide for the Collection and Preservation of Samples issued by the São Paulo State Environmental Company (CETESB, 2011), and a checklist was carried out to confirm that all materials/equipment were in accordance with the rules.

Sampling was carried out on 15 May 2013, following the pre-established points. The physical, chemical, microbiological and hydrobiological parameters of the water in the Quineira stream were analysed along the watercourse, as well as the appropriate preservation in accordance with CETESB (2011). In the field analyses, all the equipment used was calibrated beforehand to ensure that the data was reliable and accurate. The qualitative analysis of the phytoplankton was carried out using binocular optical microscopes Olympus - model CX40, and Carl Zeiss - model Primo Star, under a glass slide and coverslip. The microalgae and cyanobacteria were illustrated and photographed under a 40X objective, representing an image magnified 400 times. When necessary, the illustrated taxa were identified by comparing them with those found in specialised literature: Bicudo and Menezes (2006), Prescott et al. (1975, 1977, 1981, 1982), Bourrelly (1985), SanfAnna et al. (2006), and others. As for taxonomic classification, Round's system (1965; 1971) used by Bicudo and Menezes (2006) was adopted.

The taxa were identified superficially by their morphological characters observed under the microscope. The quantitative analysis was carried out in accordance with CETESB Technical Standard L 5.303, in a Sedgwick-Rafter chamber (1 mL) under a binocular optical microscope (10X objective) whose density was represented by n° organisms/ml - opting to count the entire chamber.

The physical, chemical, microbiological and hydrobiological parameters analysed in this study are described respectively in the following tables.

Table 01: Analysis of physical and chemical parameters in the Quineira stream micro-basin.

PHYSICAL AND CHEMICAL	UNIT	EQUIPMENT/METHODOLOGY
Air temp.	°C	Mercury thermometer
Water temp.		*M* Multisound Oximeter
Ph		pH meter
Dissolved Oxygen	mg/L	Oximeter
Electrical conductivity	µS/cm	Conductivity meter
Turbidity	UNT	Turbidity meter
Colour	Apparent	Titrimetric
Flow rate	m^3/s	Micro scooter
BOD	mg/L	BOD (5.20)
Hardness	mg CaCO3/L	Titrimetric
Alkalinity		Titrimetric
Total Phosphorus (P)		Potassium Persufate
Iron		Flame atomic absorption spectrophotometer (FAAS)
Manganese		
Zinc		
Nitrate	mg/L	
Nitrite		
Chloride		
Phosphate		Ion exchange chromatograph
Sulphate		
Fluoride		

Source: Lima, et. al (2014).

The organisms considered were isolated, colonial or filamentous taxa, and the physical, chemical, microbiological and hydrobiological parameters of the water were also analysed, as shown in the table below.

Table 02: Analysis of microbiological parameters in the Quineira stream micro-basin.

MICROBIOLOGICAL	UNIT	CULTURE MEDIUM
Escherichia coli	MPN/100 mL	Chromogenic/fluorogenic enzyme substrate
Total Coliforms		Colilert (IDEXX®)
Enterococcus sp.	CFU/mL	*Enterococcosel Agar* - BD Company
Sulphite-reducing clostridia		*Clostridium Difficile Agar* - Acumedia
Pseudomonas aeruginosa		*Cetramid Agar* - Acumedia
Heterotrophic bacteria		*Plate Count Agar* - Himedia

Source: Lima, et. al (2014).

Third stage: Identifying and proposing measures to preserve and characterise the springs of the Quineira stream.

To identify the springs, an observation was initially made along the Quineira stream to make sure there were any springs other than the two already registered, identified by SAAE (Chapada dos Guimarães Water Supply and Sewage Service). All the springs identified were photographed and described, mainly in terms of their behaviour throughout the year, and how they behave during seasonal periods. The second period was reserved for the dry months (August and September) of 2013 and 2014, when monitoring was carried out on the springs and watercourses previously identified, in order to verify their perenniality and the existence of ephemeral streams. The springs in the course of the water body were characterised according to the type of vegetation, soil and whether they were preserved or not. Monitoring was then carried out throughout 2014 to check the behaviour and classification of these springs, whether they were perennial or intermittent. The springs were monitored every three months using photographic records to analyse the behaviour of the springs found along the watercourse. This was followed by characterisation of the springs, using a camera, clipboard, pen, GPS, location map, and observation of the springs' behaviour throughout 2013 and 2014. After characterising the springs, the data was analysed and measures were proposed to preserve the springs.

CHAPTER 5

Results and Discussion

Environmental diagnosis of the Quineira micro-basin: As presented in Chapter IV of the methodology, the Quineira stream micro-basin was diagnosed in relation to the themes recommended by the VERAH method, involving the characterisation of vegetation, erosion processes, the way waste is disposed of, the quality of water resources (water), and the form of housing. Figure 04 shows the micro-basin of the Quineira stream, using the VERAH method, identified by the points: Pl (Spring), P2 (Catchment), P3 (Residences), and P4 (Outflow).

Vegetation plays a very important role, such as soil protection, as it reduces the impact of rain, protects the soil from erosive processes, contributes to increased infiltration, thus providing groundwater recharge. It acts as a filter, helping to conserve water by preventing sediment from reaching the watercourse and degrading the soil, especially for the preservation of springs. The floristic survey made it possible to observe the presence of tree, shrub and herbaceous vegetation, classified respectively as savannah, dirty field and clean field, as well as areas with exposed soil.

Figure 04: Figure illustrating the sampling points in the Quineira stream micro-basin.

Source: The author (2014).

The vegetation currently found in the micro-basin of the Quineira stream is only gallery forest and dirty fields, as shown in (figure 5). The Cerrado region that used to be so abundant in Chapada is now being urbanised. According to figure 06, which represents the map of the vegetation dynamics of the Quineira stream micro-basin, it shows how urbanisation and the coverage of this vegetation have behaved over the years, being compared every five years, according to the image in the years 1980, and 1985 are with a poorly preserved vegetation cover.

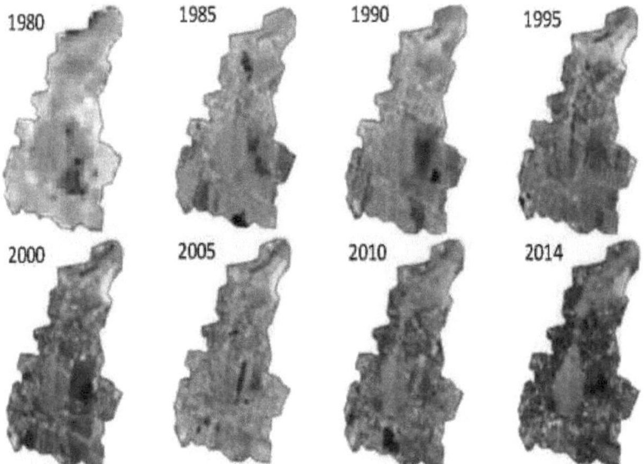

Figure 05: Vegetation dynamics in the Córrego Quineira watershed between 1980 and 2014.
Source: The author (2014).

Analysing the figure above, it can be seen that, from 1995 onwards, the vegetation is more conserved, whereas in previous years this area was more degraded. This is due to the practice of deforestation to create pastures. However, urbanisation is growing in a disorganised way, with several buildings being built along the stream and around the watershed in 2014.

According to the Brazilian Forest Code Law 12.651 of 25 May 2012, bodies of water up to 10 metres wide must have 30 metres of APP on each side. With this in mind, the APP area of the Quineira Stream is 165,764.4 metres^2 , about 4.5% of the total watershed.

Although this law establishes the minimum legal limit for Permanent Preservation Areas (PPAs), it also allows municipalities to determine higher values through the Master Plan and Land Use Laws for urban micro-basins. With 552,569.8 m^2 , or 14.99% of the watershed, this area has 33.7% of vegetation removed or altered.

Therefore, the degree of anthropogenic intervention in this area is even greater than in the legal APP, most of which is the result of urban sprawl. With the vegetation collection, it was possible to identify 23 families, 27 genera and 106 species, classified according to Vidal (2003) in all areas. The list of taxonomically differentiated species is according to LORENZI (2002), with their respective scientific and popular names, the families they belong to and the number of individuals found during the collection.

The predominant families were: Melastomataceae, Fabaceae and Cecropiaceae, these species are typical of the Cerrado. In P3, exotic plants were diagnosed: Mango (Mangifera indica L), Cashew (Anacardium occidentale), Jackfruit (Artocarpus heterophyllus), Jabuticaba (Plinia trunciflora), Banana (Musa spp), showing that urbanisation is suppressing the gallery forest in the study area, making room for exotic vegetation (Table 01).

In some residences it is more critical, as the residents have passed out to make their homes. The table below shows the variety of exotic species present on the site, most of them fruit trees.

Table 01: List of exotic species found in the Quineira Stream.

Family	Scientific Name	Popular Name
Anacardiaceae	*Mangifera indica L*	Hose
	Anacardium occidentale	Cashew tree
Moraceae	*Artocarpus heterophyllus*	Jaqueira
Myrtaceae	*Plinia trunciflora*	Jaboticaba tree
Musaceae	*Musa spp.*	Banana tree
Poaceae	-	Grasses, grasses, grasses or grasses.

Source: The author (2014).

Table 01 shows the indices (number of species and individuals) obtained at each sampling point after the laboratory processes. At PI, a total of 21 individuals were found, distributed among 11 species, the most representative being *Ingá nobilis and Trichilia catiguá,* with 6 and 3 individuals respectively. In P2, *Miconia* was the most abundant species (10 individuals), followed by *Ocotea aciphylla and Protium heptaphyllum* (both with 5 individuals). A total of 12 native species were identified, distributed among 38 plants. In P3, with 9 tree species and 18 individuals, the species are well distributed throughout the collection environment, with *Cecropia hololeuca* having the greatest abundance, with 4 individuals. In P4, the most abundant species was *Tibouchinia* granulosa with 11 individuals. In all, there were 29 individuals and only 8 species. Of the four points sampled in the survey, P3 had the lowest number of individuals (18 plants) and the second lowest richness (9 tree species), second only to P4, with 8 tree species. However, P2 stood out as the most abundant environment, with 38 individuals distributed among 12 species.

Table 02 shows the vegetation species identified in the Córrego Quineira watershed.

Table 02: List of species found in the Quineira stream watershed.

Family	Scientific Name	P1	P2	P3	P4
Anarcadiaceae	*Spondias lutea,* L.	0	0	1	0
	Tapirina guianensis, Aubl.	0	0	0	2
Annonaceae	*Xylopia sericea, St. Hil.*	1	2	0	0

Family	Species				
Boraginaceae	*Cordia brasiliensis*, (I.M.Johnst.) Gottsb.&J.S.Mill.	0	1	0	0
Burmanniaceae	*Protium heptaphyllum*, (Aubl.) March.	0	5	0	0
Cbrysobalanaceae	*Hirtella glandulosa*, Spreng.	0	2	0	0
Cecropiaceae	*Cecropia pachystachya Trec.*	0	0	4	6
Combretaceae	*Terminalia brasiliensis (A. St.-Hil.) Eichler*	1	0	0	0
Euphorbiaceae	*Mabea pohliana*, (Benth.) *Mull. Arg.*	0	0	2	3
Flacoutiaceae	*Caseara silvestre, Sw. Var. lingua*	0	2	0	1
Guttiferae	*Rheedia brasiliensis*, (Mart.) pl.et Tr.	0	0	0	1
Lauraceae	*Ocotea suaveolens, Hassl*	1	5	0	0
Leguminosae	*Anadenanthera colubrina, (v. cebil) Bren.*	0	0	1	0
	Acacia paniculata, Willd.	0	0	2	0
Malpighyaceae	*Byrsonima crassifolia*, (L.) H.B.K	0	0	0	2
Melastomataceae	*Miconia prasina, (Sw.) DC. Cogn*	2	10	0	0
	Tibouchinia granulosa, (Desr.) Cogn.	0	0	2	11
Meliaceae	*Trichilia catigua*, A. Juss.	3	2	0	0
Mimosaceae	*Ingánobilis, (Nobillis), Willd.*	6	4	2	0
	Gania sp.	0	0	3	0
Moraceae	*Brosimum gaudichaudii*, Trecul	1	0	0	0
Myristicaceae	*Virola sirinamensis, (Rol. ex Rottb.) Warb.*	2	3	0	0
Myristicaceae	*Virola sirinamensis, (Rol. ex Rottb.) Warb.*	0	3	0	0
Myrtaceae	*Myrcia rostrata, O. Berg.*	2	0	0	0
Piperaceae	*Piper tuberculatum*, Jacq.	0	0	1	0
Rubiaceae	*Duroia macrophylla*, Huber.	1	2	0	0
Sapotaceae	*Pouteria glomerata* (Miq.) Radlk	1	0	0	0
Siparunaceae	*Siparuma guianensis*, Aubl.	0	1	0	3
Number of species		**11**	**13**	**9**	**8**
Number of individuals		**32**	**55**	**27**	**37**

Source: The author (2014).

Field observations showed that the Cerrado that once covered the city is now being replaced by urbanisation. This can be seen in the land use and occupation map of the watershed (figure 07), where urbanisation has transformed the landscape to give rise to new homes. The vegetated part of the watershed is small, predominantly in the catchment area where there is a small gallery forest, which guarantees the preservation of the springs present in the watercourse.

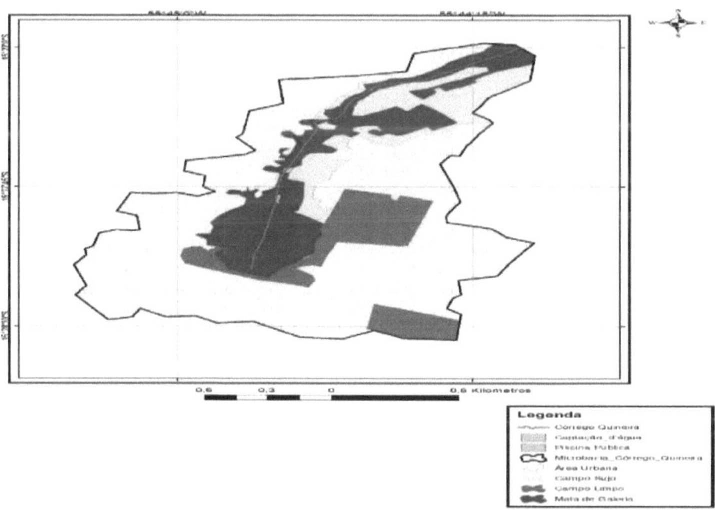

Figure 06: Land use and occupation map of the Quineira stream micro-basin.
Source: The author (2015).

Erosion

The conditions in which these soils occur allow them to function specifically in relation to the dynamics of the physical environment. The Latossolos, being naturally permeable (Salomão, 1994), favour the infiltration of rainwater, while the Plintossolos, due to their shallow subsurface drainage impediment layer made up of ferruginous armour, hinder the infiltration of water, favouring surface runoff and, consequently, erosion. The high permeability of latosols is due to the porous structure observed in the surface and subsurface horizons, regardless of their texture, making them low erodibility soils (Salomão, 2007).

Five linear erosion events were identified in the Quineira stream catchment, two of which were representative diagnoses. The registered erosions were named "Boçoroca do Complexo de Nascentes" and "Erosão do Sr Nadir". The "Boçoroca do Complexo de Nascentes" erosion is located in the deep valley, more specifically in the area of the springs of the Córrego Quineira, which is classified in Geotechnical Unit UG4, has a length of approximately 300 metres, a maximum depth of 2.6 m and an average width of 6 m, totalling a volume of 4680 m^3 . In this case, undisciplined rainwater from the urban area, mainly due to the waterproofing of the land and the streets, drained down the talvegue line at the bottom of the valley. This initially led to the formation of furrows, which progressed to form gullies and then intercepted the groundwater aquifer. This led to the formation of the phenomenon known as *piping* (underground erosion), characterising a gully (figure 07).

Figure 07: Gully showing the piping phenomenon.
Source: The author (2014).

It was found that the gully is in the process of evolving due to the various evidence of *piping* phenomena within it. In addition, there is still concentrated surface water flowing into it from the urban area, forming small furrows.

Figure 08: Erosion in the street in front of the houses.

Source: The author (2014).

Mr Nadir's Erosion" is located on Rua Maria Martins da Paixão, a small dirt road (40m), in a sloping area at the transition from the geotechnical unit of the plateau top (UG1) to that of the deep valley (UG4). It is a linear furrow-type erosion 30 m long, with a maximum depth and average width of 0.4 m, totalling a volume of 4.8 m^3 (Fig. 10). The furrows developed as a result of the concentrated flow of rainwater from the asphalted roads in the upper part of the area, with a slope of around 15% (gently undulating terrain). Over time, it became a vehicular passageway to give access to the homes (3 so far) that have been irregularly installed in an area unsuitable for habitation because it is located within the Quineira basin area.

Solid waste and housing

According to information collected in the field, solid waste collection and cleaning of public roads in the municipality of Chapada dos Guimarães/MT is carried out during the day (morning and afternoon) from 7am to 11am and from 1pm to 5pm. Actions to prevent, regulate and supervise environmental damage, as well as the collection and final disposal of solid waste are the responsibility of the Town Hall, through the Municipal Department of Tourism and the Environment and the Chapada dos Guimarães Department of Works and Public

22

Services. The waste is sent to the dump, which is approximately 12 km from the centre of Chapada dos Guimarães, 9 km of which are paved roads and 3 km of which are unpaved.

According to Vieira et al (2011), the structure of the municipal Solid Waste Management system, with the implementation of the Landfill and the Waste Treatment and Composting Plant at Km 3 of the MT-20 motorway, would be implemented by December 2013, along with the selective solid waste collection system. Figure 10 shows the area where waste from the city of Chapada dos Guimarães is disposed of, an open site without any treatment and with a lot of accumulated waste.

Figure 09: Area where rubbish is dumped by residents of Chapada dos Guimarães/MT.
Source: The author (2014).

In the area around the source and catchment, solid waste was found that contains various types of plastic, such as pet bottles, disposable cups, bags, envelopes from industrialised foodstuffs, etc. Below plastics came metals: cans, metal paper lunch boxes and bottle caps. Paper waste came third: cardboard, boxes, packaging and loose paper in general, and lastly glass.

In the pool area there is a large amount of irregularly disposed waste, figure 10 shows a pet bottle dumped in the area, this material takes a long time to decompose, a large concentration of plastic bags was also observed in the area; and to a lesser extent construction materials, aluminium cans and paper, which can contribute to the degradation of water resources.

Fonte: O autor, (2014).

Figura 10:   The figure shows some waste in the catchment of the Quineira stream.

23 interviews were carried out in the commercial sector. The questions were related to the solid urban waste generated in the establishments, final disposal, as well as rubbish bins, collection, storage, recycling, among others. According to the results of the interviews, 59 per cent of the establishments do not carry out selective collection, and the other 41 per cent that do, only separate dry materials from wet ones.

When asked about the quality of waste collection, 73% of respondents considered waste collection to be good, 23% said it was regular and 4% terrible, as shown in figure 13, but always emphasised that the quality of the service offered by the town hall needed to be improved.

Qualidade da coleta

Figura 11:   Graphical representation of establishments in terms of collection quality.
Source: The author (2014).

Most of the waste collected from households (97%) is collected directly by the cleaning service and 3% is disposed of in skips for later collection, as shown in Figure 11.

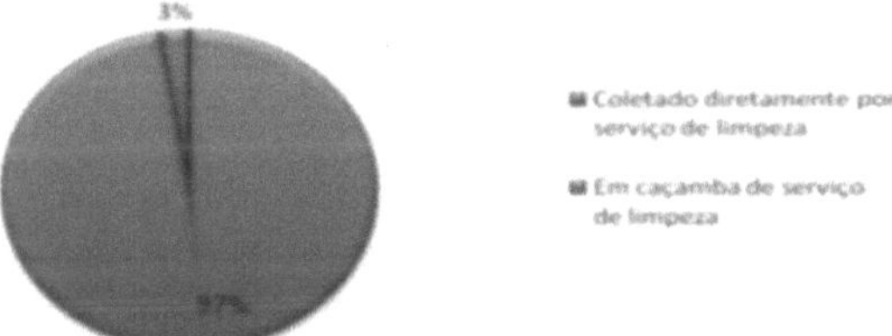

Figure 12: Graphical representation of establishments regarding the final destination of waste.
Source: The author (2014).

The areas with the highest concentration of solid waste are represented in Figure 12 with red hatches and the regions where the questionnaires were carried out are identified with red dots.

Hospitality establishments, whether inns or hotels, had a capacity of between 40 and 90 people. It was also noted that the monthly flow of accommodation varies according to the time of events that take place in the town, such as the "Winter Festival", which is organised in the town. The number of toilets in establishments of this type varies according to their size and capacity, with the largest establishment having 32 toilets and the smallest 9 toilets. In the other establishments, the number of toilets varies between two and three, with one used for staff and the other two for customers.

Figure 3 shows, in yellow, the area where the housing questionnaires were administered. The light blue mark refers to the place where the interviews were carried out in the town's shops.

Figura 13: The figure shows where the questionnaire was administered in the watershed.
Source: The author (2014).

Another problem is diffuse pollution, which is difficult to identify as it does not have a specific point of discharge, has very different characteristics and occurs throughout the basin (GRILLI and BETTINE, 2010). During rainfall, the washing of streets and roofs can carry numerous contaminants into the stream. In addition, improperly installed septic tanks used for sewage disposal can seriously compromise water quality. According to Bonilha et al. (2008), the municipality of Chapada dos Guimarães does not have a sewage treatment system, and each household is responsible for its own treatment system using simple pits, i.e. wells dug into the ground for the purpose of sewage infiltration.

Water

The sampling points established for collecting water samples to analyse and assess the quality of the physical, chemical, microbiological and hydrobiological parameters were carried out from the source to the outlet of the Quineira stream, as shown in (figure 1), and are called P1, P2, P3, P4, P5 and P6: P1 - Spring Complex, P2 - Confluence of Springs, P3 - SAAE Catchment, P4 - Public Swimming Pool, P5 - Residence, P6. The diatom fructules and dinophycean tecarn plates were unable to receive clarification and dehydration treatment, or even the preparation of permanent slides, so the taxa were identified superficially by their morphological characters observed under the microscope. After interpreting the physical, chemical, microbiological and hydrobiological results, it can be said that both the spring complex and its confluence are relatively well preserved. Although the results of the bacteriological analyses were outside the standards established by current legislation at points 1 and 2, it cannot be said that the anthropisation around the

26

watershed is affecting the quality of the water at these sampling points.

Figure 14: Water collection sites in the Córrego do Quineira watershed.
Source: The author (2014).

Given that the data point to natural conditions, this can be confirmed by the lack of rainfall, which connects the surroundings to the bodies of water through surface runoff, where various residues and organic materials are carried away. It can be seen that along the longitudinal gradient of the Quineira stream up to point 4, there are various changes in the parameters assessed, mainly due to the characteristics found such as the type of soil, changes in the substrate, as well as factors such as the damming of one of the sampling points for collection, treatment and subsequent distribution for public supply.

However, downstream of sample point 4, domestic sewage was found *on site on* both banks of the streams, through clandestine connections to the rainwater galleries, showing that the alterations that occurred at points 5 and 6 are strongly linked to anthropogenic actions in the watershed that is the subject of this study. Table 3 shows the values obtained for the physical and chemical parameters in the Córrego Quineira watershed in Chapada dos Guimarães.

Table 03: Results of the physical and chemical parameters found in the Quineira stream micro-basin.

PHYSICAL AND CHEMICAL	P1	P2	P3	P4	P5	P6	Reference standard[1]
Water temperature (° C)	23,5	23,4	23,5	-	22	-	-
PH	5,39	5,56	5,88	6,14	7,02	7,35	6.0-9.0
Dissolved Oxygen (mg.L $)^{-1}$	8,10	8,20	5,00	7,20	6,80	8,24	>5
Electrical conductivity (µS.cm $)^{-1}$	6,02	8,62	12,85	22,80	41,80	58,20	-
Turbidity (NTU)	0,12	0,41	0,12	0,18	0,13	0,14	100
Colour (uH)	7	87	15	42	13	19	75
Flow (L.s $)^{-1}$	3	41	-	35	132	95	-
BOD (mg.L $)^{-1}$	1,6	2,5	3,20	1,20	1,60	2,90	<5
Hardness (mgCaCOs.L $)^{-1}$	0	0	0	0	115,14	59,47	-

Alkalinity (mgCaCOs.L)$^{-1}$	-	18,0	5,0	-	-	23,0	-
Total Phosphorus (mg.L)$^{-1}$	0,015	0,035	0,042	0,020	0,017	0,019	3

Source: Lima, et. al (2014).

CONAMA Resolution 357/2005 establishes a limit of not less than 5.0 mg/L of dissolved DO for Class 2 rivers. The recovery of dissolved DO in the water body is linked to the aeration of the system (waterfalls, etc.) and natural self-depuration processes, which occurs at the other sampling points. Looking at the results obtained in the field, all the points are within the standards established by legislation. However, it was observed that the level of dissolved oxygen decreases at point 03, probably due to the formation of a lentic environment as a result of the water intake dam.

CHAPTER 6

CHARACTERISATION OF THE QUINEIRA STREAM SPRINGS

Figure 15 shows the delimitation of the Quineira stream micro-basin and the places where the springs occur, identified by points 1, 2, 3, 4, 5, 6 and 7.

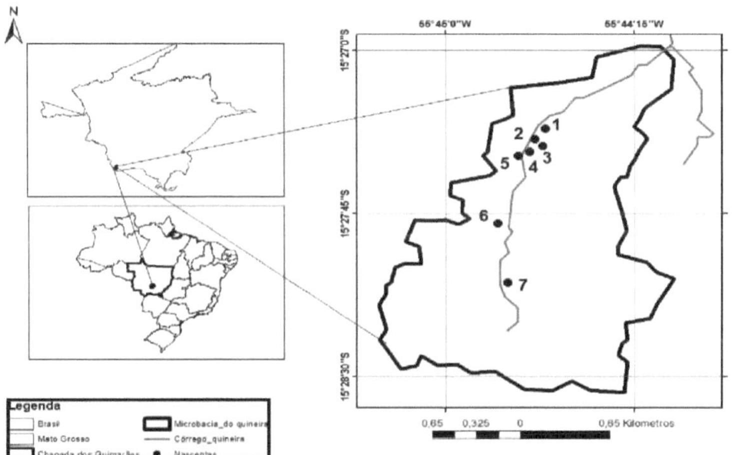

Figura 15:   Delimitation of the Quineira stream micro-basin and location of the springs.

Source: The author (2014).

Figure 16 illustrates the dammed area where the water catchment station for water supply is located, in May 2013, inside the Quineira park. The area was completely flooded, which contributes to increasing the water supply for the town of Chapada.

Figura 16:   Catchment of the Quineira stream, during the dry season.

Source: The author (2014).

The springs occur in fine sandstone rock, the material above the rock is a colluvium, 1.20m of armour, with alteration of the same, all are perennial that manifests itself throughout the year, but with flows varying throughout the same, are partially degraded, with the presence of little vegetation around them, with location along the watercourse, and above is the gallery forest.

Spring 3 shows a lot of sediment brought in by the torrent, especially during rainy periods, while springs 4, 5, 6 and 7 are partially degraded. Andrade (2005) states that the preservation of springs is of fundamental importance, as damage to them affects not only the Quineira Stream, but also the surrounding streams, which already face problems with sewage and have undergone a recovery process.

Figure 17: Shows the springs of the Quineira stream.

Source: The author (2013).

With the creation of the Chapada dos Guimarães Municipal Park, the springs are better protected, but with the population growth that is suppressing the micro-basin of the Córrego Quineira, water resources are jeopardised. The environmental function of the APP has been partially maintained, since the APP has a medium degree of degradation and this compromises the maintenance of water resources, since the construction of new homes compromises the existing vegetation on the site, the removal of this vegetation can cause problems related to soil protection, the maintenance of biodiversity and the landscape, which is in total degradation. According to Crepani et al (2001), as well as protecting the soil, vegetation prevents erosion and landslides.

Figure 21 A, B, C, D and E shows the vegetation along the watercourse, the site is an area of gallery forest with tree species and small shrubs. Despite being an APP area, there is construction work less than 50 metres from the stream, which contributes to the degradation of the springs in the stream. Figure F is spring 7, where the water that supplies the town is collected. Figure 21 F shows the rocks in more detail

| Nasc. 6 | | Nasc. 7 |

The waterfall is surrounded by gallery forest vegetation, but there is no vegetation around the spring. Its banks are partially preserved, but when it rains there is a large amount of rubbish in the vicinity, which causes silting on the banks of the stream.Riparian forests are of the utmost importance for the quality of river water, preventing water regime control, reducing erosion on river banks, maintaining fauna by increasing the supply of fish, and improving the landscape (FERREIRA apud BOTELHO, 2006). Figure 22 shows the course of the Quineira stream and its riparian forest, which is suffering from erosion and a large amount of solid waste on its banks.

Figure 18: Shows the course of the Quineira stream.

(A) (B) (C) (D) (E) (F)

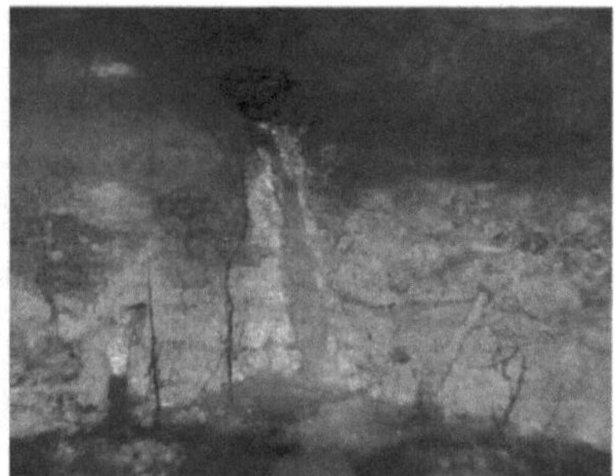

Source: The author (2013).

According to Araújo (2002), APPs are areas in which, by law, the vegetation must be kept intact, with a view to guaranteeing the preservation of water resources, geological stability, biodiversity *and* ensuring the well-being of human populations. The protection regime for APPs is quite rigid: the rule is untouchability, with the exception of the removal of the original vegetation cover only in cases of public utility or legally established social interests. Exposed soil is more susceptible to erosion, ravines and gullies. In the case of the Quineira stream, practically the entire bank of the stream is suffering from erosion, a factor related to the lack of vegetation around it.

When recovering the vegetation cover of APPs that have already been degraded, a distinction must be made between the guidelines for the type of water upwelling, i.e. without or with initial water accumulation, because waterlogging of the soil or the temporary submergence of the plant root system, the depth of the profile and the fertility of the soil are some of the factors that must be considered, as they are selective for the species that will be able to develop (RODRIGUES; SHEPHERD, 2000).

Finnotti et al (2009) states that improperly disposed urban waste is leached by precipitation and can contribute to the load of pollutants that are directed into rivers. Another type of impact demonstrated by the author concerns the removal of vegetation from urban watersheds in general for occupation. Taking into account the importance that these areas represent for maintaining the quality and quantity of water resources, the impact of this removal is quite significant. During the month of July, the water level is very low, thus jeopardising the city's water supply. The information was provided by

by SAA, pumps from the Monjolo and Samambaia streams are switched on during this period so that the supply is not jeopardised and the population does not run out of water.

According to Philippi Jr (2008), the urban environment is characterised by being an ecosystem where changes are more significant, leading to highly altered characteristics. Among the main characteristics of this environment, the author highlights the following: High population density; Disproportionate relationship between the built environment and the natural environment; Importation of energy to keep the system running; High volume of waste; Significant alteration of native biological diversity, with the removal of forests and the importation of plant and animal species. The imbalance of the main biogeochemical cycles, such as the water, carbon, nitrogen and phosphorus cycles; soil sealing and the alteration of watercourses.

CHAPTER 7

Final considerations

The environmental diagnosis carried out in the Quineira stream micro-basin by applying the VERAH Method and identifying and characterising the springs made it possible to understand the current state of degradation of surface water resources, in order to support the necessary actions aimed at minimising impacts and preserving the watercourse, with the following conclusions standing out:

The vegetation currently found in the Quineira stream micro-basin is only gallery forest and dirty fields. There is a small area of gallery forest near the springs, which partially preserves the watercourse. Five linear erosion processes were found, two of which are representative because they are within one of the main springs, which was determined to be a gully and is in the process of evolving due to the various evidence of *piping* phenomena. With regard to waste, there is a large amount deposited in the Quineira watershed and along the watercourse, which directly harms the preservation of surface water resources and damages the springs, thus compromising water quality. This is due to the disorganised construction of housing around the watershed; the deposit of waste has become much greater, both construction and domestic, which totally compromises the preservation of the springs.

With the physical, chemical and microbiological results of the water analysis, it can be said that both the spring complex and its confluence are relatively unspoilt. All the results are within the standards required by the Conarna Resolution, which ensures the quality of water resources. The characterisation of the springs shows that they are in a compromised state of conservation, i.e. partially degraded.

The lack of vegetation around these springs jeopardises the functioning of the water resources, as well as leaving the soil totally susceptible to erosion processes, and compromises the quality of the water because of the amount of rubbish that falls on the banks of the stream, especially during rainy seasons. Parque do Quineira is an APP area, with springs along the watercourse, which should be reserved so that they don't suffer from degradation processes.

CHAPTER 8

BIBLIOGRAPHY

ANDRADE, Ana Lúcia. Community defends lagoons: residents clean up the water, but come up against the destruction of springs and predatory fishing. A Tarde, Salvador, p.7, 30 Aug. 2005.

ARAÚJO, S. M. V.G. de. **Permanent preservation areas and the urban question.** Legislative Consultancy of the Chamber of Deputies, Brasília, Aug. 2002. Available at: Accessed on: 26 Jan. 2015.

BARROS, F. G. N. AMIN, Mário. M. **Agua: um bem económico de valor para o Brasil e o mundo.** Revista Brasileira de Gestão e Desenvolvimento Regional, Jan-Apr/2008, Taubaté, SP, Brazil. (ARTICLE).

BRAZIL, **Brazilian Forest Code.** Law No. 4.771, of 15 September 1965.

BRAZIL. Law No. 12.651 of 25 May 2012. **Provides for the protection of** native **vegetation**; amends Laws Nos. 6.938, of 31 August 1981, 9.393, of 19 December 1996, and 11.428, of 22 December 2006; repeals Laws Nos. 4.771, of 15 September 1965, and 7.754, of 14 April 1989, and Provisional Measure No. 2.166-67, of 24 August 2001; and makes other provisions. Brasília: DOU, 28/05/2012.

BRAZIL. CONAMA Resolution No. 357 of 17 March 2005. Provides for the **classification of water bodies and environmental guidelines for their classification.**

BRAZIL. Ministry of Mines and Energy, General Secretary. RADAMBRASIL Project. Sheet SE.21 Corumbá and part of sheet SE.20; **geology, geomorphology, pedology, vegetation and potential land use.** Rio de Janeiro. 452 p. ill., 5 maps (Levantamento de Recursos Naturais, 27).

BICUDO, C.E.M, M. **Gêneros de Algas de águas Continentais Brasileiras: chave para identificações e descrições.** 2 ed. São Carlos: RIMA, 2006. Pag. 502. Bonilha A. R. et al, 2008. PRE-PLAN OF THE MONJOLO STREAM BASIN IN CHAPADA DOS GUIMARÃES - MT. Postgraduate Programme in Water Resources/Federal University of Mato Grosso.

BONILHA, A. R.; MAGALHÃES, A.; SILVA, C.; BELIQUE, E.; SILVEIRA, I. S. F.; SANTOS, L. G.; BENECCIUTI, L.; ROCCHI, M.; GARCIA, S. S.; SILVA, V. J.;
SILVA, W. C.; **Pre-plan of the Monjolo stream basin in Chapada dos Guimarães -MT. Cuiabá- MT.** 2008.

BRUCE, N. **Introduction to Environmental Chemistry.** Winnipeg: Wuerz Publishing Ltd, 1993.

CASTRO JÚNIOR, Prudêncio Rodrigues de. **Soil Erosion.** Cuiabá: Instituto Pró- Natura, 2002.

CALHEIROS, Rinaldo de Oliveira, **Preservação e Recuperação das Nascentes.** et.al. Piracicaba: Comité das Bacias Hidrográficas dos Rios CJ - CTRN, 2004.

CALHEIROS, Rinaldo de Oliveira, et. al. **Caderno** de **Mata Ciliar. State Secretariat for the Environment, Department for the Protection of Biodiversity.** - N 1 (2009)--São Paulo: SMA, 2009.

CALIJURI, Maria do Carmo; BUBEL, Anna Paola Michelano. **Conceptualising micro-watersheds.** In : LIMA, Walter de Paula, and ZAKIA, Maria José Brito. Planted Forests and Water - Implementing the Concept of the Microwatershed as a Planning Unit. São Carlos - RiMa, 2006. pg.45-59.

CETESB - Companhia de Tecnologia e Saneamento Ambiental. **Determination of the most probable number of sulphite-reducing clostridia (clostridium perfringens)**: test method. Technical Standard L5.213 São Paulo: CETESB, 1993.

CETESB - São Paulo State Environmental Technology and Sanitation Company. **National sampling guide: water, sediment, aquatic communities and liquid effluents.** Organisers: Carlos Jesus Brandão...(et al.) - São Paulo: CETESB; Brasília: ANA, 2011.

CETESB - Companhia de Tecnologia e Saneamento Ambiental. **Determination of the most probable number of sulphite-reducing clostridia (clostridium perfringens): test method.** Technical Standard L5.213 São Paulo: CETESB, 1993.

CETESB - São Paulo State Environmental Technology and Sanitation Company. **National sampling guide: water, sediment, aquatic communities and liquid effluents.** Organisers: Carlos Jesus Brandão, (et al.) - São Paulo: CETESB; Brasília: ANA, 2011.

CREPANI, E.; Medeiros, J. S. De; Hernandez Filho, P.; Florenzano, T. G.; Duarte, V.; Barbosa, C. C. F. **Remote Sensing and Geoprocessing Applied to Economic Ecological Zoning and Territorial Planning.** São José dos Campos, 2001 (INPE-RPQ/722).

CRISTOFOLETTI, Antônio. Geomorphology. 2 d. 9ª reprint (2006). São Paulo: Edgard Blucher, 1980.

COUTINHO, L.M. 1978. **The concept of Cerrado. Revista Brasileira de Botânica** 7:17-23.

COUTINHO, L.M. 2002. **The cerrado biome. In Eugen Warming e o cerrado brasileiro:** um século depois (A.L. Klein, ed.). Editora da Unesp, São Paulo, p.77-91.

DICTIONARY, Ecological Environmental, available at: http://www.ecolnews.com.br/dicionarioambiental/. Accessed on: 12 October 2013.

FARIA, A. P. (1997). **"The dynamics of springs and the influence on** channel **flows".** A Água em Revista, Ri o de Janeiro, v. 8, pp. 74-80.

FINOTTI, Alexandra R. et al. **Monitoring water resources in urban areas.** Caxias do Sul: Educas, 2009. 272 p.

FELIPPE, M. F. **Characterisation and typology of springs in conservation units in Belo Horizonte (MG) based on geomorphological, hydrological and environmental variables.** Master's thesis - Federal University of Minas Gerais, Belo Horizonte, 2009.

FERREIRA, W.C. **Establishment of Riparian Forest in Degraded and Disturbed Areas.** Lavras, Minas Gerais: UFLA, 2006. Apud DAVID, A.C.; CARVALHO, L.M.T.; BOTELHO, S.A. **Identification of areas with potential for natural regeneration around the Funil HPP reservoir.** Lavras: CEMAC/UFLA 2003. 352p. (Technical Report).

GRILLI, M.; BETTINE, S. C. (2010) **Estimation of diffuse pollution in the Santa Cândida farm stream basin.** In: Proceedings of the XV Meeting of Scientific Initiation at PUC-Campinas - 26 and 27 October.

GUERRA, A. J. T.; MARÇAL, M. S. Geomorfologia Ambiental. Rio de Janeiro: Bertrand do Brasil, 3ª ed., 2006.

GUERRA, Antonio José Teixeira; ALMEIDA, Josimar Ribeiro de; ARAÚJO, Gustavo Henrique de Souza. **Environmental Management of Degraded Areas.** Editora Bertrand Brasil, 2005.
Geological Glossary/IBGE, **Department of Natural Resources and Environmental Studies.** - Rio de Janeiro: IBGE, 1999.

GLOSSARY, International Hydrological : http://webworld.unesco.org/water/ihp/db/glossary/glu/PT/GF1185PT.HTM. Accessed on: 12 October 2013.

LAW N⁰ 12.651, OF 25 MAY 2012. Available at: http://www.ipef.br/pcsn/documentos/lei12651.pdf. Accessed on: 12 October 2013.

LORENZI, H. Árvores Brasileiras: **Manual de identificação e cultivos de plantas arbóreas do Brasil.** 2. ed. São Paulo: Nova Odessa, 2002.

LAW No. 12.651, OF 25 MAY 2012, **NEW FOREST CODE Provides for the protection of native vegetation.**

MAITELLI, G. T. Hydrography. In: MORENO, G. & HIGA, T. C. S. **Geografia de Mato Grosso.** Cuiabá, Entrelinhas. 2005.

OLIVEIRA, Antônio Manoel dos Santos. VERAH - **Environmental Diagnosis of Microbasins.** Cuiabá: UFMT, 2008.

PRESCOTT, G.W., Croasdale, H.T. Bicudo, C. E. M. & Vinyard , W.C. 1982. **A synopsis of North American desmids.** Part II: Desmidiaceae: Placodermae. Section 4. University of Nebraska Press, Lincoln, London.

PRESCOTT, G.W., Croasdale, H.T. Vinyard , W.C. & Bicudo, C. E. M. 1981. **A synopsis of North American desmids.** Part II: Desmidiaceae: Placodermae. Section 3. University of Nebraska Press, Lincoln, London.

PRESCOTT, G.W., Croasdale, H.T. & Vinyard, W.C. 1977. **A synopsis of North American desmids.** Part II: Desmidiaceae: Placodermae. Section 2. University of Nebraska Press, Lincoln, London.

PILICIONI; Andréa Focesi Pelicioni **I Trajectory of the Environmental Movement.** In: Curso de gestão ambiental, pg 19. ed. Manole, 2004.

PHILLIPI JUNIOR, Arlindo. **Sanitation, health and the environment: foundations for sustainable development.** 2. ed. São Paulo: Manole Ltda, 2008. 842 p.

CONAMA RESOLUTION No. 357 of 17 March 2005. **Provides for the classification of water bodies and environmental guidelines for their classification.**

CONAMA RESOLUTION no. 274 of 29 November 2000. **Defines the criteria for bathing in Brazilian waters.**

CONAMA RESOLUTION No. 004, of 18 September 1985. The National Environment Council - CONAMA, using the powers conferred on it by Law No. 6.938, of 31 August 1981, Decree No. 88.351, of 1 June 1983, amended by Decree No. 91.305, of 3 June 1985, Decree No. 89.336, of 31 January 1984, and in view of what is established *by* Law No.⁰ 4.771, of 15 September 1965, amended by Law No. 6.535, of 15 June 1978, and by what is determined by CONAMA Resolution 8/84.

CONAMA RESOLUTION, RDC no. 275, of 21 October 2002. Provides for the Technical Regulation of Standard Operating Procedures applied to Food Producing/Industrialising Establishments and the Checklist of Good Manufacturing Practices in Food Producing/Industrialising Establishments. D.O.U. - Diário Oficial da União; Poder Executivo, 23 October 2003.

RODRIGUES, R.R.; SHEPHERD, G. **Conditioning factors of riparian vegetation.** In: RODRIGUES, R.R.; LEITÃO FILHO, H. F. (Eds.). **Riparian forests: conservation and recovery.** São Paulo: USP/FAPESP, 2000. chap. 6. p. 101-107.

SALOMÃO, F. X. T. Linear erosion processes in Bauru-SP. **Cartographic regionalisation applied to preventive urban-rural erosion control. 1994.** Thesis (Doctorate) - Faculty of Philosophy, Letters and Human Sciences, University of São Paulo, São Paulo.

SALOMÃO, F. X. T. **Control and prevention of erosion processes.** In: GUERRA, A. J. T.; SILVA, A. S.;

BOTELHO, R. G. N. **Erosão e conservação dos solos: conceitos, temas e aplicações.** Rio de Janeiro: Editora Bertrand do Brasil, 3ª ed., 2007. p. 229267.

SCHUMM, S. A.; HARVEY M. D.; WATSON C. C. Incised Channels. Water Resources Publications, Littleton, Colorado.

SCHREINER, Simone. Climate and altitude in tropical cities: the example of Chapada dos Guimarães and a comparison with Cuiabá-MT. Master's thesis - Federal University of Mato Grosso, Institute of Human and Social Sciences, Postgraduate Programme in Geography, 2009.

STRECK, E. V.: **Environmental education for the conservation and recovery of the** environment. Porto Alegre: EMATER/R. Available at: Shttp://www.ana.gov.br/; BAESA, 2007. 28p.

SALOMÃO, Fernando Ximenes de Tavares; MADRUGA, Elder de Lucena & MIGLIORINI, Renato Blat. **Geotechnical map of the urban perimeter of Chapada dos**

Guimarães: subsidies for the masterplan. Geol. USP, Sér. cient., São Paulo, v. 12, n. 1, p. -15, April 2012. SALOMÃO, F. X. T. **Control and prevention of erosion processes.** In: GUERRA, A. J. T.; SILVA, A. S.; BOTELHO, R.G. **Erosão e conservação dos solos: conceitos, temas e aplicações.** Rio de Janeiro: Bertrand Brasil, 1999, p.229-267.

SALOMÃO, F. X. T. **Control and prevention of erosion processes.** In: GUERRA, A. J. T.; SILVA, A. S.; BOTELHO, R. G. N. **Erosão e conservação dos solos: conceitos, temas e aplicações.** Rio de Janeiro: Editora Bertrand do Brasil, 3ª ed., 2007. p. 229267.

SALOMÃO, F. X. T. *Linear erosion processes in Bauru-SP. Cartographic regionalisation applied to the* preventive *control of urban-rural erosion.* 1994. Thesis (Doctorate) - Faculty of Philosophy, Letters and Human Sciences, University of São Paulo, São Paulo.

SILVA, Ricardo Toledo. **Water** Resources **and Urban Development** . In: MUNOZ, Héctor Raúl. **Interfaces da Gestão de Recursos Hídricos:** desafios da lei de Águas de 1997.2 ed. Brasília, Secretaria de Recursos Hídricos, 2000. p. 280-293.

SILVA, Ricardo Toledo. **Water Resources and Urban Development.** In: MUNOZ, Héctor Raúl. **Interfaces da Gestão de Recursos Hídricos:** desafios da lei de Águas de 1997.2 ed. Brasília, Secretaria de Recursos Hídricos, 2000. p. 280-293.

SILVEIRA, A.L.L. **Hydrological cycle and river basin.** In: TUCCI, C.E.M. (Org.). Hidrologia: ciência e aplicação. São Paulo: EDUSP, 2001. p 35-51.

SOUZA. C.G., et al. **Caracterização e manejo integrado de bacias hidrográficas.** Belo Horizonte: EMATER, 2002. 124p.

TODD, David K; MAYS, L. W. (2005). **Groundwater hydrology**. 3rd ed. Wiley, Hoboken, NJ.

TORRES, R. A.; JUNQUEIRA, F. J. A. L. **Increasing milk productivity and quality in the Zona da Mata Mineira - Juiz de Fora:** Embrapa Gado de leite, 2005. Chap. 9. p.103-111.

TUNDISSI, J.G. Limnology in the 21st Century: **Perspectives and Challenges**. International Institute of Limnology. São Carlos, SP, 24 p, 1999.

TUNDISI, J. G. **The Future of Resources,** October 2003. Available at: http://www.multiciencia.unicamp.br/artigos 01/A3 Tundisi port.PDF.

Accessed: 05 July 2014.

TUCCI, Carlos E. M. Hidrologia: **Ciência e Aplicação**, (Organiser), 3ª Ed., first reprint. - Porto Alegre, Editora da UFRGS/EDUSP/ABRH, 2004, pag, 40.

URBANA, In: Hidrologia: Ciência e Aplicação, Tucci, C.E.M. (Organiser), Editora da UFRGS/EDUSP/ABRH, 1993.

TUCCI, C.E.M. **Hidrologia: ciência e aplicação. 3ª edition, Porto Alegre:** Ed. Da UFRGS / ABRH, 2003. Pg, 22.

URBANA, In: Hidrologia: **Ciência e Aplicação**, Tucci, C.E.M. (Organiser), Editora da UFRGS/EDUSP/ABRH, 1993.

VALENTE, Osvaldo F.; GOMES, Marcos A. (2005). **Conservation of** springs: **hydrology and management of headwater catchments.** Aprenda Fácil, Viçosa, Pg, 28.

VALENTE, Osvaldo F. GOMES, Marcos Antônio. **Conservation of Springs: Water Production in Small Watersheds.** Viçosa - MG: Aprenda Fácil, 2011, pg. 111, 112, 114, 115, 117, 118, 119.

VIEIRA, JUNIOR, Hamilcar Tavares. **Geopark Chapada dos Guimarães-MT:** proposal. SIG Mapa / Hamilcar Tavares Vieira Junior; Juliana Maceira Moraes; Carlos Schobbenhaus. Goiânia: CPRM, 2011 (Geoparks Project) 6, 60 f.: il.

Vidal, Waldomiro Nunes. Vidal, Maria Rosária Rodrigues. **Botanical Organography; illustrated synoptic tables of phanerogams.** 4º ed. ver. Ampl. - Viçosa: UFV, 2003, Pg, [22.]

I want morebooks!

Buy your books fast and straightforward online - at one of world's fastest growing online book stores! Environmentally sound due to Print-on-Demand technologies.

Buy your books online at
www.morebooks.shop

Kaufen Sie Ihre Bücher schnell und unkompliziert online – auf einer der am schnellsten wachsenden Buchhandelsplattformen weltweit! Dank Print-On-Demand umwelt- und ressourcenschonend produziert.

Bücher schneller online kaufen
www.morebooks.shop

Printed by Books on Demand GmbH, Norderstedt / Germany